Energy and Electromagnetism

Developed at
The Lawrence Hall of Science,
University of California, Berkeley
Published and distributed by
Delta Education,
a member of the School Specialty Family

1325246
978-1-60902-039-2
Printing 3 — 10/2012
Quad/Graphics, Versailles, KY

Table of Contents

Edison Sees the Light

"The **filament** burns out too quickly," Mr. Edison said. "We have to find a better material to make a longer-lasting filament."

Thomas Edison (1847–1931) was the most famous inventor of his time. He invented the phonograph, the motion-picture camera, the first copy machine, and hundreds of other things. He is most famous, however, for improving a product he *didn't* invent, the electric **lightbulb**.

The problem with lightbulbs before 1879 was that they burned out too quickly. The filament is the part of the lightbulb that actually makes the light. When an **electric current** flows through the filament, the filament gets so hot that it glows and gives off light. The hotter the filament gets, the brighter the light. But the hotter the filament gets, the faster it burns out.

Edison with his lightbulb

Edison's short-lived lightbulb was a simple device. It was much like a modern incandescent lightbulb. In an incandescent lightbulb, two stiff support **wires** hold the filament. A clear glass globe surrounds the filament for protection. The glass globe is attached to a metal casing. The tricky part is how the support wires, which are part of the **circuit**, connect to the metal casing.

One support wire attaches to the side of the metal case. The other support wire attaches to a small metal disc at the bottom of the base. The base **contact point** must not touch the main part of the metal case. This is important. When **electricity** travels to the lightbulb in a circuit, the electricity must flow *through* the filament.

When you put a lightbulb in a circuit, electricity can be delivered to the lightbulb. When the circuit is complete, the electric current will flow. The electric current has **energy**. The energy produces **heat** and **light** as the lightbulb does its job. Energy leaves the lightbulb system as light.

Edison tackled the filament problem with hard work. He is credited with saying, "Invention is 1 percent inspiration and 99 percent perspiration." Edison directed his team to try every imaginable material to find the best filament. It is said that they tried and rejected 2,000 materials. Edison needed help.

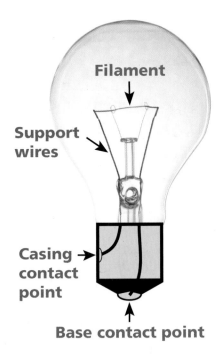

Filament

Support wires

Casing → contact point

Base contact point

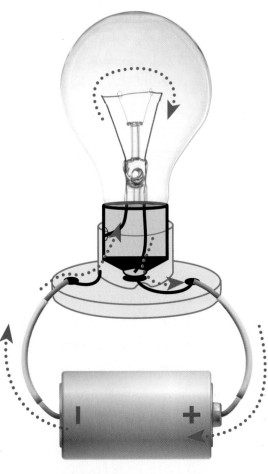

A lightbulb in a circuit

Help came in the form of Lewis Latimer (1848–1928). Latimer was an experienced draftsman and inventor. He had been working on the filament problem, too. Latimer discovered that a carbon-coated cotton thread made a good filament. He got a patent for the carbon filament. Inventors get patents from the government when they invent something new. When Edison tried the carbon filament in his lab, he agreed that it was the best material. Edison bought the patent from Latimer. Now Edison could use the carbon filament in his lightbulb.

Lewis Latimer

Edison had to solve one more problem to make a useful lightbulb. He knew that things need oxygen to burn. He predicted that if he could remove the air from the glass globe, there would be no oxygen, and the filament would not burn up. He was right. This new lightbulb lasted months instead of days.

Thomas Edison had seen the light. Now it was time to show this new light source to the world. It was New Year's Eve in 1879. Edison's team strung lights from their lab to the train station. A crowd of more than 3,000 people came to see what would happen.

It was a very dark night, and all the gaslights had been turned off. Edison stepped up to the platform and threw the switch. All the lights came on. The crowd cheered.

Edison understood the importance of electric lighting. It could change the American way of life. That's why he asked Latimer to join his team in 1884. Latimer stayed with Edison for years. He wrote patents for new inventions and books on electrical engineering.

Edison's lab

Many years later, in 1918, the team of scientists and engineers gathered to celebrate Edison's birthday. They called themselves the Edison Pioneers. Lewis Latimer was the only African American among the engineers. He also was one of the 28 founding Pioneers.

Lightbulbs Today

In Edison's time, the only way known to make electric light was to make a filament so hot that it glowed. The glowing filament gave off a lot of heat and a good amount of light. It takes a lot of energy to make light by heating a filament. Today we have alternative ways to make light that don't need nearly as much energy.

The long white tubes that produce light are called fluorescent lamps. A fluorescent lamp does not have a filament. Instead the tube is filled with gas. When an electric current travels to the lightbulb, the gas begins to glow and give off light. The light is not quite as bright as an incandescent lamp. But the amount of energy needed to produce the light is far less than the energy needed to heat a filament.

The Edison Pioneers in 1920

We also have compact fluorescent lightbulbs. The tube is much thinner, and it is wound into a coil to save space. Compact fluorescent lightbulbs screw into standard sockets designed for incandescent lightbulbs. Replacing all of your incandescent lightbulbs with compact fluorescent lightbulbs can save a household several hundred dollars every year.

In 1962, a new light-producing technology was developed. It was a tiny device called a light-emitting diode (LED). LEDs produce light by using a small amount of energy to emit a ray of light. The LED doesn't waste energy by producing heat. The first LEDs were dim and produced only red light. But they were extremely efficient.

As electrical engineers continued to develop new LEDs, they developed amber- and green-colored LEDs. The colored lights made it possible to convert traffic lights to LEDs. This saved cities a lot of money. Eventually, an LED was developed that produced pure, bright white light. The newest technology for lighting homes and businesses is LED lighting because modern LEDs can produce bright white light using much less electricity, resulting in huge cost savings. You might have seen flashlights that use clusters of small bright lights instead of a single lightbulb. Those small bright lights are modern LEDs.

Thinking about Lightbulbs

1. How do you know when energy is moving in a lightbulb circuit?

2. Describe the path taken by electricity through an incandescent lightbulb.

3. What are some of the reasons why lamp technology has changed?

Modern LEDs

Electricity

Energy makes things happen. Every action is caused by energy. For example, energy makes things warm. Energy makes things move. Energy makes sound and light. Energy is everywhere, and it makes things happen.

There are many ways we can observe energy at work. Some of those ways are heat, **motion**, **sound**, and light. Most of the energy we use comes from the Sun. Energy comes from the Sun as light and heat. Light and heat can make things happen. Think about standing in the sunshine. You can see the light and feel the heat.

The Sun constantly radiates energy that reaches Earth.

A toaster plugged into a wall socket

Listening to a portable music player powered by a battery

Energy Sources

Electricity is used to make hundreds of different things happen. Electricity can make light. Electricity can make sound. Electricity can make things move. Electricity can make things hot or cold.

Many electric appliances, such as toasters and lamps, have a cord with a plug on the end. They are plugged into a wall socket. The socket is connected to a wire. The wire is connected to a **generator** at a power plant many miles away. You are using the electricity as it is being generated. A lamp uses electric current to produce light. A toaster uses electric current to produce heat. As long as your lamp or toaster is connected to the **energy source**, it will do its work.

But what if you want to take your music player with you to the beach? There is no place to plug in a music player at the beach. But you can still listen to music. You just need a **battery**. A battery is a portable source of **stored energy**.

Stored Energy

A battery is a source of stored energy. Batteries are full of chemicals. The chemicals in a battery can produce electricity. Electricity from batteries is the same as electricity from a wall socket, but portable. Electricity makes many things happen. Batteries can make sound in a music player and light in a flashlight. The stored energy in batteries can also start a car, power a cell phone, or drive a toy boat across a pond.

Batteries come in all sizes. Hearing aids that fit inside a person's ear have tiny batteries. Batteries that provide the energy to start and drive cars and buses are large. Hybrid vehicles that run on both an electric engine and a gasoline engine have lots of batteries to serve different functions.

Hearing aids are powered by stored energy in tiny batteries.

A car battery

Huge batteries power this electric school bus.

Some communities have electric buses, even electric school buses. The batteries take up most of the space under the seats. The energy in these large batteries flows through wires as electric current. The current powers electric **motors**. The motors turn the bus wheels to move the bus. This bus never has to go to the gas station.

Electric cars are starting to show up on city streets and highways around the United States. They never stop for gas because they need a different source of energy for refueling. Electric cars need to have their batteries recharged. Recharging a battery requires connecting the battery to a source of electricity. Batteries produce electricity when the chemicals in them react. The reaction forms new chemicals and produces electricity. When the chemicals have all reacted, they stop producing electricity. By connecting the battery to an energy source, the chemicals are changed back into their starting conditions. The recharged battery is again ready to produce electricity as a product of the chemical reaction in the battery.

An electric car is recharged by connecting it to an energy source.

So how is an electric car recharged? The driver parks the car near an energy source. The driver plugs an extension cord from an energy source into the car. While the car sits there, the batteries recharge. Recharging can happen at home or at a recharging station. Recharging can also happen at work or at a shopping center.

Electricity is the most popular way to transfer or move energy to get a job done. Producing electricity involves changing an energy source into electric current. Scientists are looking for alternative sources of energy to produce electricity. Some alternative sources of energy are light and heat directly from the Sun, moving air or wind, heat generated inside Earth, and moving water.

Questions to Explore

1. What is energy?

2. What can electricity do?

3. What is a battery?

Energy

Burning Fuel

Energy is present in many different places. Energy comes from an energy source. One source of energy is fuel. Fuel is material that has stored energy. People burn fuel to release the stored energy. How can you tell that energy is released when fuel burns? You can see the light and feel heat. Wherever there is light and heat, there is energy.

Candle wax is a fuel. When candle wax burns, the energy of the wax can be observed as heat and light. Coal is a fuel. When coal burns, the stored energy of coal is released as heat. Burning coal produces heat to boil water in a steam train. The steam from the boiling water turns the train wheels. The energy of coal puts the steam train into motion. Wherever there is an object in motion, there is energy.

Natural gas is another kind of fuel. When gas burns, the energy of the gas produces heat. Heat from the burning gas can do many things. Can you think of some ways to use this heat?

Candle wax is an energy source.

Coal is the fuel that makes steam to power this train.

Burning natural gas makes heat.

Cars, trucks, and buses use fuel to move on a road.

Oil, gasoline, and wood are also fuels. Burning oil releases heat. People use heat to warm homes and make electricity.

Burning gasoline releases energy to make cars and trucks move. The energy of oil and gasoline produces heat and motion.

People burn wood to provide heat for homes. And heat from burning wood in a campfire can cook food.

Heat from a campfire cooks food.

Food is the fuel that provides energy for people.

Food Is Fuel

A slice of pizza or a piece of fruit tastes good. Pizza and fruit are examples of food. But do people eat food only for the taste? No, food is the fuel that makes life happen.

Food has stored energy, just like other kinds of fuel. Living organisms "burn" fuel to release energy. The food doesn't really burn with a flame, like wood or coal. Animals digest food, and the stored energy is released.

Animals, including people, use food to produce motion and heat. Sled dogs digest food to provide energy to run, pull the sled, and keep warm. What do you do with the energy of food?

All animals eat food to get the energy they need to live.

Food gives you energy to play sports.

15

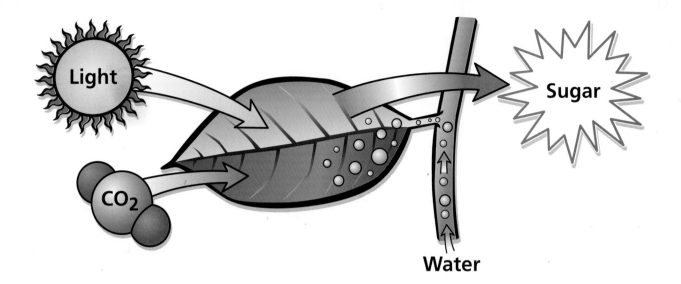

How does energy get into food? You might be surprised to know that the energy of food comes from the Sun.

Light is one way that energy moves. Light from the Sun shines on plant leaves. The leaves absorb energy.

The leaves take up water from the soil and carbon dioxide (CO_2) gas from the air. Plants make sugar from the CO_2, water, and energy of the Sun's light. Sugar is food.

Energy stored in food is released when you eat and digest food. So really, it is energy from the Sun that keeps you warm, lets you run, and makes all your other activities possible.

Food has stored energy.

It takes energy to stand, point, look, and think.

Motion

Did you transfer any energy today? The answer is yes. Every second of every day, you are transferring energy. The action of lifting a pencil takes energy. The action of looking at a picture takes energy. The action of thinking about what to eat for lunch takes energy. Every action requires energy.

The energy for lifting, looking, moving, and thinking comes from food. The energy of food is stored in chemicals. As food breaks down in your body, energy produces heat and motion, and even electric currents. Electric currents send messages throughout your brain and along your nerves. Heat, movement, and electric impulses are important ways that energy moves to keep you and your friends alive.

Making Sound

Sound is another way to observe that energy is present. When objects vibrate, they produce sound. A **vibration** is a fast back-and-forth movement. A vibrating object pushes on air. That back-and-forth motion pushes on the air and makes waves that create sound.

The sound waves travel through air. They carry the energy created by the vibrating (moving) object. The sound waves hit your ear. The energy carried by the sound waves moves the little bones in your ear. You hear the sound.

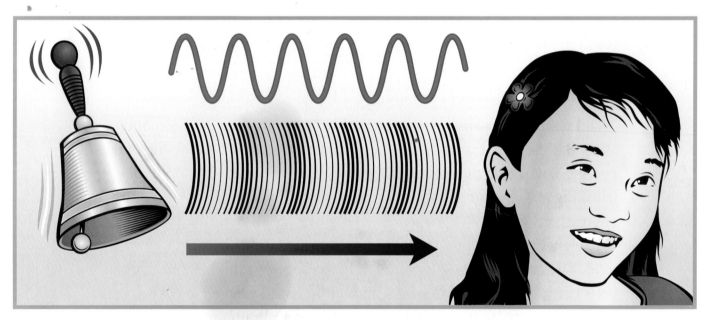

Waves carry sound from a bell to your ear.

A fern fossil

Fossil Fuels

Do you know what a fossil is? You may have seen fossils of leaves, shells, or bones. They are usually found in layers of rock. Fossils are the preserved remains of organisms that lived long ago. We know that dinosaurs lived on Earth 65 million years ago. That's because we have fossils of dinosaur bones that we can study.

Oil, coal, and natural gas are called **fossil fuels**. Scientists think that fossil fuels started as organisms that lived a long time ago. For millions of years, these organisms died and piled up on Earth's surface. Over time they got buried deep underground. Slowly, the organisms changed into oil, coal, and gas. That's why they are called fossil fuels. They are the ancient remains of organisms.

Fossil fuels are made of chemicals. The chemicals in fossil fuels can be burned to produce useful energy. Burning fossil fuels produces heat.

How do we use fossil fuels? People everywhere use oil for transportation. Cars, buses, trucks, ships, trains, and airplanes burn fuels made from oil to make them move.

Coal is used a lot in the midwestern and eastern United States. Coal is burned to generate heat and electricity. Coal is also burned to make steel.

In the western part of the United States, people use a lot of natural gas to generate electricity. The burning gas heats water to make steam. The steam turns generators to make the electricity used in homes and schools.

Oil is a fossil fuel.

Coal is a fossil fuel.

Natural gas is a fossil fuel.

Observable Evidence of Energy

Stored energy is useful to people when it produces heat, light, sound, and electric current, and puts objects in motion. Heat, light, sound, electric current, and objects in motion are evidence that energy is present and available to do work.

The energy of food produces heat and motion in our bodies when digested.

The energy of fuel produces heat and light, and can sometimes produce motion.

The energy of batteries produces electric currents in a **complete circuit**. Electric currents can produce heat, light, motion, and sound.

Look at the table below. The left-hand column lists sources of stored energy. The right-hand column lists the evidence of energy the source can produce.

Stored-energy source	Produces observable evidence of energy
corn	heat and motion
turkey	heat and motion
carrot	heat and motion
wood	heat and light
wax	heat and light
oil	heat and light
natural gas	heat and light
coal	heat and light
battery	heat, light, sound, and motion

Thinking about Energy

1. What evidence shows that energy is present?

2. How are food, fuel, and batteries alike?

3. What is the source of most of the energy used by people?

Series and Parallel Circuits

Electric current comes from an energy source. The source might be a D-cell battery, a solar cell, or a wall socket. When a **component**, like a lightbulb, is connected to a source of electricity, the lightbulb will make light. When a different component, like a motor, is connected to an electricity source, the motor shaft will turn. How do you connect a lightbulb or a motor to an electricity source?

You can use a D-cell to light a lightbulb. Metal wires carry the electricity. If you try to get the lightbulb to light using one wire like this, the lightbulb will not shine.

An incomplete circuit

The trick is to use two wires. One wire connects the base of the lightbulb to one end of the D-cell. The second wire connects the metal casing of the lightbulb to the other end of the D-cell. This setup results in a bright, shining lightbulb. It is called a complete circuit, or a **closed circuit**. The places on a D-cell and lightbulb where wires touch the component are called contact points.

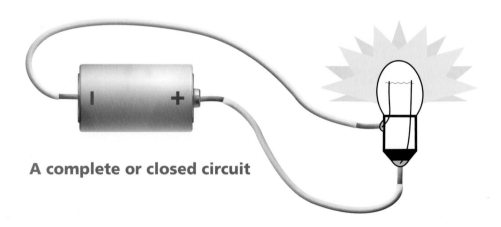

A complete or closed circuit

If you disconnect one of the wires from the lightbulb or from the D-cell, the lightbulb will stop shining. This is because the pathway through which the electric current flows to the lightbulb is broken. A circuit with a break is called an **incomplete circuit**, or an **open circuit**.

It is important where the wires connect to the D-cell and the lightbulb. One wire must connect to the positive (+) end of the cell. The other wire must connect to the negative (–) end of the cell. The other end of one of the wires must connect to the metal casing of the lightbulb. The other end of the second wire must connect to the base of the lightbulb. These connections make a closed circuit. The electric current will flow, and the lightbulb will shine.

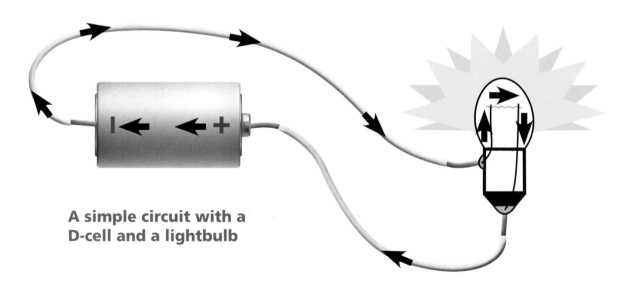

A simple circuit with a D-cell and a lightbulb

You might want to connect two lightbulbs to a D-cell. How can you do this? There are two ways. You can open the one-lightbulb circuit and put a second lightbulb into the circuit. Now the electric current flows through two lightbulbs in one circuit. This is a **series circuit**. There are two lightbulbs and one D-cell connected in series. In a series circuit, current has only one pathway to flow from the source (D-cell) to the components (lightbulbs).

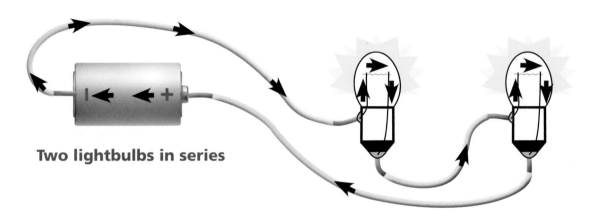

Two lightbulbs in series

There is another way to add a second lightbulb to the one-lightbulb circuit. You can use two wires to connect the second lightbulb to the first lightbulb. This is called a **parallel circuit**. There are two lightbulbs in parallel connected to a D-cell. In this parallel circuit, each component (lightbulb) has its own pathway to the energy source (D-cell). Lightbulb 1 is in the blue-arrow pathway; lightbulb 2 is in the red-arrow pathway.

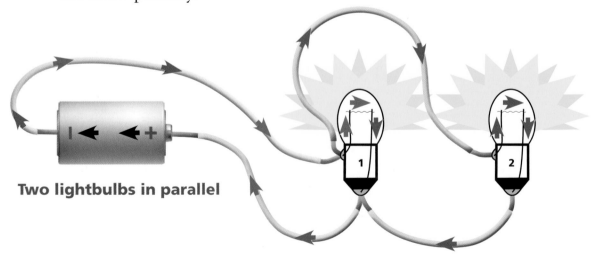

Two lightbulbs in parallel

Which Circuit Should You Use?

Suppose you want to light two lightbulbs. Is there any reason to put them in parallel rather than in series? Yes. The reason becomes clear when you compare the two kinds of circuits. Two lightbulbs in series both shine with a dim light. Both lightbulbs in parallel shine brightly. If you want bright lights, put the lightbulbs in parallel.

Why is there a difference? Two lightbulbs in series have to share the energy of the D-cell. There is only one pathway for the electric current. The current flows from the negative end of the cell through the first lightbulb. It then goes through the second lightbulb and back to the positive end of the cell.

Two lightbulbs in parallel do not have to share energy of the D-cell. Each lightbulb has its own pathway to the source of electricity. Even though some wires are shared for part of the pathway, more than one pathway lets each lightbulb get its own electricity. That's why lightbulbs in parallel shine more brightly.

So, is it better to connect your lightbulbs in series or in parallel? It seems like parallel would be better because you get two bright lights. But there is a cost. The energy of the D-cell will drain much faster when it is supplying electricity to two lightbulbs in parallel. When lightbulbs are connected in series, the D-cell lasts longer, but the lights are dimmer.

Two or more lights connected in parallel shine brightly.

Adding More D-Cells to a Circuit

If you want to put two or more D-cells in a circuit, they can be connected in series or in parallel. Which of these circuits shows two cells in series? Which shows two cells in parallel?

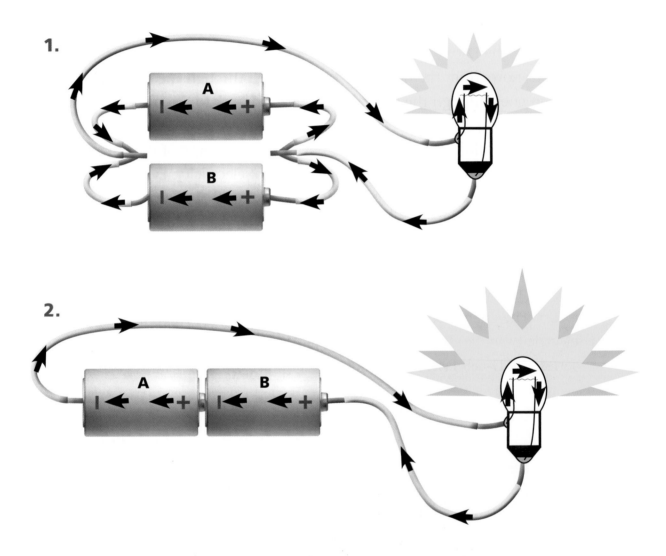

In circuit 1, cells A and B are in parallel. Each cell delivers electricity to the lightbulb in its own pathway. In circuit 2, electricity is delivered to the lightbulb by two cells working together in the same pathway. As a result, the lightbulb in circuit 2 will be twice as bright as the lightbulb in circuit 1.

Here is an interesting circuit made with lightbulbs and D-cells. How would you describe this circuit?

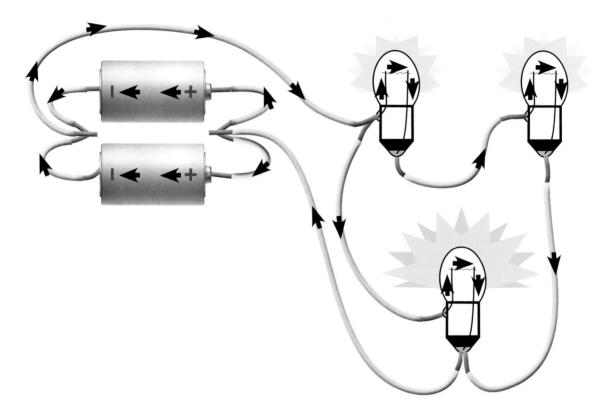

This circuit is one lightbulb in parallel with two lightbulbs in series, powered by two D-cells in parallel. If you said that, you got it right.

Thinking about Circuits

1. What is the advantage of wiring two lightbulbs in parallel?

2. Why are two lightbulbs in series with a D-cell dim?

3. Do you think the lights in your home are wired in series or in parallel? Why do you think so?

A house generating electric energy with solar panels

Alternative Sources of Electricity

People use electricity. We use it to light classrooms, to run computers, and to toast bread.

Most of our electricity comes from power plants that burn fossil fuels. Oil, coal, and natural gas are burned to heat water. The steam from the water moves rapidly and spins turbines. The turbines generate electricity. This electricity is expensive to produce. Burning fossil fuels is not good for the environment. And fossil fuels will not last forever.

Remember when you took a solar cell outside to run the motor? You were using an alternative energy source to produce electricity. Sunshine is a renewable resource. If you connected the motor to a battery and let it run, the battery would soon die. You would probably recycle the dead battery and buy a new one. If you set up the same motor with a solar cell, it would keep going for years. Could these little solar cells be used on a larger scale? You bet!

Let's look at a few ways to produce electricity using alternatives to fossil fuels. These natural sources of energy are renewable. Renewable resources are replaced as we use them. We will not run out of renewable resources. And using renewable resources is much better for the environment.

Solar Power

When people want to use solar energy to generate electricity, they use solar cells that are wired together to create a solar panel. Sometimes they are called photovoltaic cells. Often the solar panels are wired in series circuits. Then they are placed on the roofs of schools, businesses, and homes to generate electricity. There are even solar-cell power plants in deserts that help generate electricity for cities.

An apartment building using solar panels to generate electricity

Solar cells absorb the Sun's light and change it into electricity. Electricity is used to power lights, computers, and appliances, and even to heat our homes. If you have a lot of photovoltaic cells and a sunny day, you can create a lot of electricity. Sometimes these solar panels make more than enough electricity for the building on which they are installed. The extra electricity can be sold to power companies.

A solar power plant

Wind turbines

Wind Power

Wind is powerful. During powerful storms, the wind is strong enough to knock down big trees. On milder days, wind can move large sailboats quickly across a bay. Do we use the wind's energy to create electricity? Yes, we do.

Wind turbines on tall towers use the strong winds above Earth's surface to generate electricity. Strong and steady wind makes the long blades spin. The spinning blades turn a generator to create electricity. Wind turbines are expensive to build. But the wind is free and clean. Wind farms are being built around the country where there is strong, steady wind on most days of the year.

Geothermal Power

Some places in the world have vents leading from Earth's mantle, up through its crust, to the surface. Water can collect underground near a vent and heat up. When this happens, the steam and water shoot out in the form of geysers. Do engineers put the steam-generating geysers to work? Yes, they do.

A power plant captures Earth's heat to generate electricity. How does it do this? A power company drills into the vent until it reaches an area where the water collects. The steam that flows through the hole spins a turbine to generate electricity. The power company condenses water vapor gas back to liquid water by cooling it. The liquid water returns to the ground to get hot again and to generate more electricity.

An engineer at a geothermal
power plant

Hydroelectric Power

Have you ever seen a swollen river in the springtime after heavy rains? If you have, you probably know how powerful moving water can be. Hydroelectric power plants are built across rivers. They have turbines that spin as the water falls from a high point to a low point. The turbines are placed in the path of the falling water to generate electricity. Hydroelectric power is used across the country where rivers flow from mountains to valley floors.

Building a hydroelectric power plant across a river is not always good for the plants and animals that live there. We need to be careful about what we do to the environment. We need to make sure that the benefits of building a power plant are greater than the damage caused to the environment.

A hydroelectric power plant

Reading outside using natural light

Saving Electricity

One of the best ways to make sure that we have enough energy is for everyone to use less electricity. Energy conservation is a quick and easy way for families and schools to save money. It also helps conserve Earth's natural resources. Here are just a few of the things you can do. When you're not using a computer, turn it off. When you're leaving a room, turn the lights off. When you need something from the refrigerator, decide ahead of time what you're going to get, open the door, and get it as quickly as you can.

Maybe someday you'll have a job working with alternative sources of energy. By then, maybe they will be our main sources of electricity!

Ms. Osgood's Class Report

Our school is located on an island called Vinalhaven, in the state of Maine. If we want to go ashore to the mainland, we need to take a 12 kilometer (km) ferry ride from our home in Penobscot Bay. It is very expensive to buy electricity on the island because it travels so far underwater. We often lose power in the winter during big storms. On November 13, 2009, we went on a field trip. We went to watch the ribbon-cutting ceremony for three new wind turbines. The wind turbines will provide most of the electricity for two islands. We hope the wind turbines will make our electricity cost less.

It was very exciting when the towers and blades arrived. They came from the mainland on huge barges at high tide. None of us had ever seen towers that were so wide. Part of the excitement was watching how the workers got the parts off the barge and onto the trucks. The huge blades were transported to the site and assembled. They were lifted into the air by gigantic cranes and attached to the towers. We were surprised to see how big the turbines are up close.

Here are two of the wind turbines! Can you guess where the photographer was when he took this picture?

When Magnet Meets Magnet

Y ou know what a refrigerator **magnet** is. It's an object that might look like a flower, a piece of fruit, or a seashell. And, most importantly, it sticks to the refrigerator. Refrigerator magnets come in thousands of sizes, colors, and shapes. But they all have one thing in common. When you look on the back, you see a piece of black flexible material or a hard, black object. The flexible piece or black object that sticks to the refrigerator is a **permanent magnet**.

The magnet might look different from the doughnut-shaped one you used in class. But it works just the same. The shape on the front is a decoration. It's the magnet that sticks to the refrigerator.

The most interesting thing about magnets is that they stick to refrigerators and to a lot of other things. But they don't stick to everything. If you test all the different objects around your kitchen, you will soon discover that magnets only stick to some metal objects.

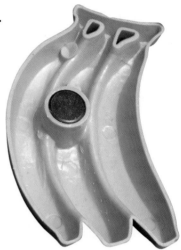

A permanent magnet on the back of a refrigerator magnet

Refrigerator magnets

Visit FOSSweb to explore more about magnets!

36

After more testing, you will find that magnets stick to one kind of metal. That metal is **iron**. Iron can be mixed with other metals to make steel. Magnets stick to steel because steel is mostly iron. Magnets do not stick to objects made of most other metals. For example, magnets do not stick to aluminum pots, copper coins, silver spoons, gold rings, or brass hinges. The general rule is *if a magnet sticks to an object, the object is iron or steel.*

When you bring two magnets together, two things can happen. Sometimes magnets pull on each other and stick together. When they pull and stick, we say they **attract**.

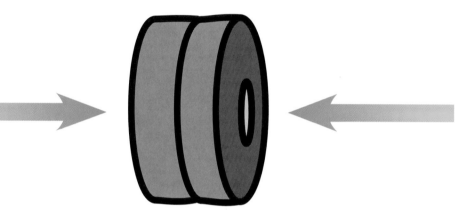

Two magnets attracting

At other times, magnets push each other apart. When they push apart, we say they **repel**. Why do you think they repel?

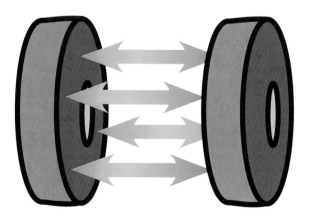

Two magnets repelling

Magnetic Poles

Every magnet has two different sides or ends called **poles**. One pole is called the **north pole**, and the other is called the **south pole**. A simple bar magnet has its two poles on opposite ends. A horseshoe magnet has a pole on each end of the horseshoe. The doughnut magnets you worked with have poles on the two flat sides. Magnets always have a north pole and a south pole.

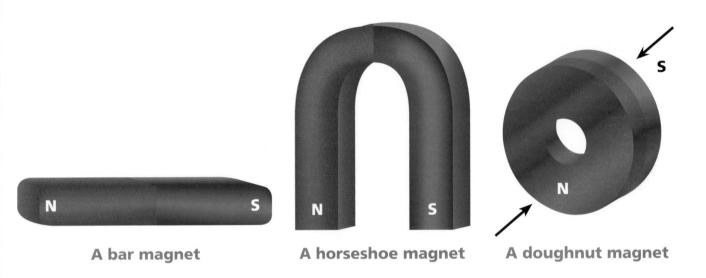

A bar magnet   A horseshoe magnet   A doughnut magnet

You might wonder what happens when a bar magnet breaks. Do you have a magnet with just one pole? No, both pieces still have a north pole and a south pole. The same is true for all other magnets. No matter how many pieces you cut a magnet into, each piece still has a north pole and a south pole.

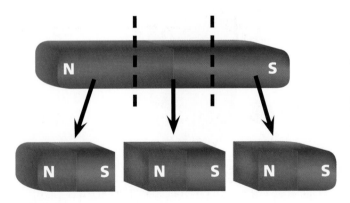

Cut a long bar magnet into three pieces. Each piece has a north pole and a south pole.

What happens when two magnets come close to each other, and you can feel them repel? How are the poles **oriented**? Do the magnets repel when two south poles come together? Do the magnets repel when two north poles come together? Or do they repel when one south pole and one north pole come together?

Here are four pairs of bar magnets being held together. Which ones will push apart when they are released?

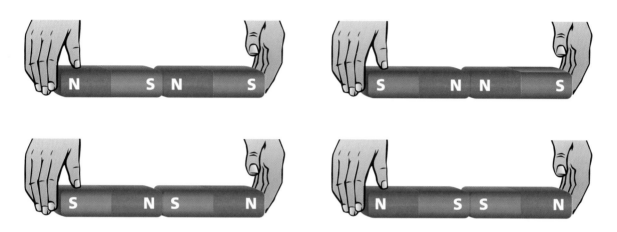

This is what happens when the magnets are released. The two pairs of magnets on the left attract each other. The two pairs of magnets on the right repel each other.

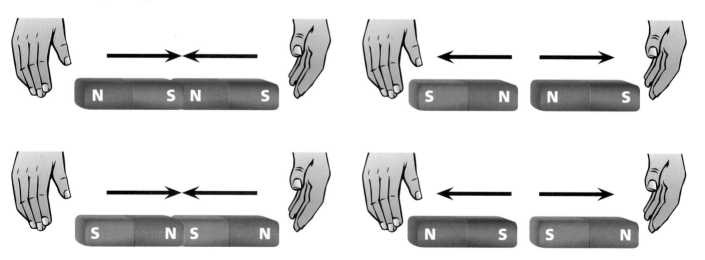

Two north poles always repel each other. Two south poles always repel each other. We can state a general rule. *Like poles repel.*

A north pole and a south pole always attract each other. It doesn't matter which magnet has the north pole and which has the south pole. We can state another general rule. *Opposite poles attract.*

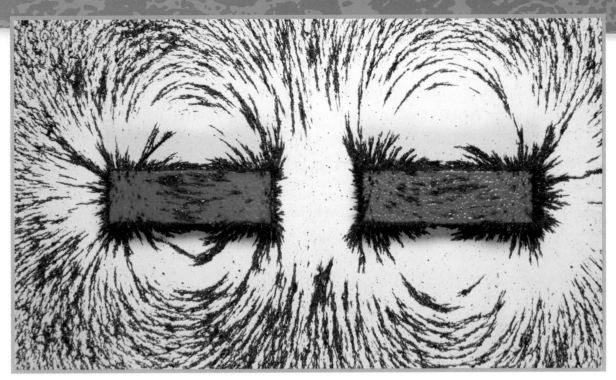

Using iron filings, you can see magnetic fields around magnets.

The Magnetic Force

Magnetism is a **force**. A force is a push or a pull. You can feel the magnetic force when you pull two attracting magnets apart. In the same way, you can feel the magnetic force when you push two repelling magnets together. The force of magnetism is what makes magnets act the ways they do.

How Magnets Stick to Iron

If opposite poles attract, why does a magnet stick to a piece of iron? Magnetism extends out from a magnet in an invisible area called a **magnetic field**. When a magnet comes close to a piece of iron, such as a steel nail, the magnetic field **interacts** with the iron in the nail. The nail becomes a **temporary magnet**. The end of the nail becomes one pole of a magnet. The magnet then sticks to the temporary magnet. So magnets don't really stick to iron. Magnets stick to other magnets.

The magnetism in the iron is called **induced magnetism**. Induced magnetism happens only when a magnet is close by. If you bring the south pole of a magnet close to the head of a nail, what pole will the head of the nail become? Just apply the rule. Opposites attract. (The nail head will become a north pole.)

The First Magnets

The first magnets were pieces of a naturally occurring mineral called magnetite. Magnetite sticks to magnets because it contains a lot of iron. The black rock in your set of test objects is magnetite. When magnetite is by itself, it is called lodestone.

Legend has it that shepherds found bits of rock sticking to the iron nails in their sandals more than 2,000 years ago. One area rich with lodestone was a part of present-day Turkey. This region was called Magnesia. The word *magnet* may come from the name of this ancient region.

Magnetite is also found in a number of locations in the United States. Some of the major sources are shown on the map below. Can you find the major magnetite source closest to where you live?

Magnetite

Lodestone was found in Magnesia.

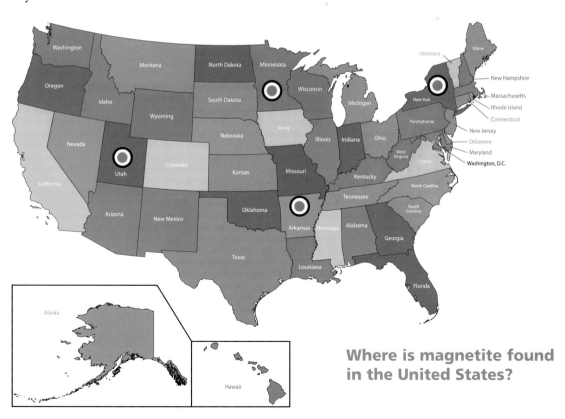

Where is magnetite found in the United States?

41

Reviewing Magnets

Magnets are objects that stick to iron and steel. All magnets have two poles, a north pole and a south pole. Like poles repel. Opposite poles attract.

Thinking about Magnetic Interactions

1. Why does magnetite stick to a magnet?

2. What causes magnets to attract each other at some times and repel each other at other times?

3. The magnets shown below have one pole labeled. Which pairs of magnets will attract, and which will repel?

Magnificent Magnetic Models

The ancient Romans and Egyptians used magnets to create artwork. They did not understand magnetic force. But they knew how to use it to do wonderful things with magnets. The Romans used magnets to suspend a figure of the god Mercury in midair in one of their temples.

The Egyptians hung iron and lodestone figures from ropes. The figures would repel and attract each other so that they appeared to dance. You can make models of these statues. Look on the next pages for plans.

Roman and Egyptian artwork using magnets

Dancing Statues

What You Need

2 Small plastic containers
14 Doughnut magnets
1 Ruler or pencil
• String
• Scissors

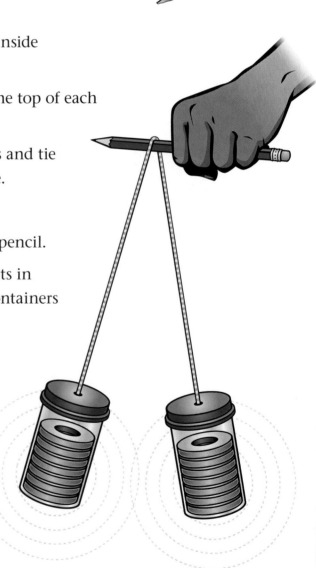

What You Do

1 Place seven doughnut magnets inside each container.

2 Use the scissors to cut a slit in the top of each container, from side to center.

3 Slide the string through the slits and tie knots to hold the string in place.

4 Put the tops on the containers.

5 Hang the string from a ruler or pencil.

6 You may have to flip the magnets in one container over if the two containers do not repel each other.

What Happens

The containers dance around each other and never touch.

The Suspended Statue

What You Need

1 Glass jar with a steel screw-on lid
1 Steel paper clip
1 Magnet
• Tape
• Thread

What You Do

1 Stick the magnet to the underside of the jar lid.

2 Tie the paper clip to one end of the thread.

3 Be sure the thread is the right length. The paper clip should almost reach the lid when the thread is taped to the bottom of the jar.

4 Tape the other end of the thread to the bottom of the jar.

5 Hold the jar upside down. Screw on the lid. Then turn the jar upright again. The paper clip should not touch the magnet.

What Happens

The paper clip floats mysteriously at the end of the thread.

Discuss Your Ideas

1. Explain why the containers filled with magnets act the way they do.

2. Why doesn't the paper clip fall to the bottom of the jar?

3. If you wanted to have three containers of magnets dance around one another, how would you orient the magnets?

Make a Magnetic Compass

You can make a **compass** just like the one a hiker might use to keep from getting lost. Here's how to do it.

What You Need

- 1 Bar magnet
- 1 Ruler
- 1 Piece of thread about 30 centimeters (cm) long
- 1 Store-bought compass
- Masking tape

What You Do

1 Tie one end of the thread to the middle of the magnet so that the magnet hangs level.

2 Tie the other end of the thread to the end of a ruler.

3 Tape the ruler at the edge of a table so that the magnet hangs in midair. Make sure the magnet is not close to any steel. (Do you know why?) Your compass is done!

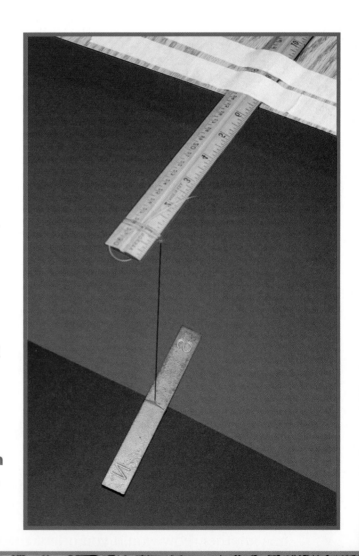

A compass made with a bar magnet, a ruler, and a piece of thread

What Happens

When the magnet comes to rest, it will point north and south because it is a compass. But which end points north? You will need a store-bought compass to find out.

The painted end of the store-bought compass needle will point north. Now you should be able to figure out which end of your bar magnet is pointing north.

You can make sure by slowly bringing the store-bought compass up to one end of your hanging bar magnet. Do the compass needle and the magnet point in the same direction? If the answer is yes, then both magnets are pointing north.

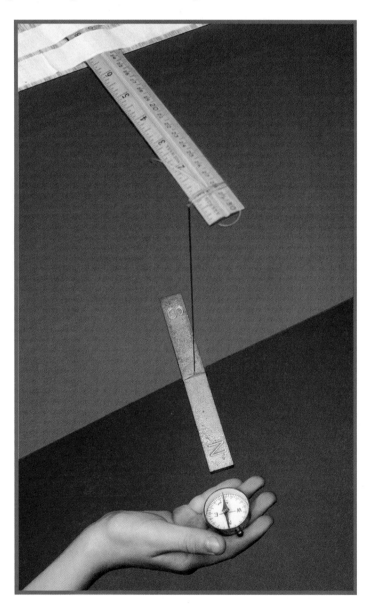

A store-bought compass next to a homemade compass

Magnetic Fields

Every magnet has a magnetic field around it. Think of a magnetic field as many invisible lines that loop from the north pole to the south pole of a magnet, and then through the magnet. When two magnets interact, it is actually the magnetic fields of the two magnets that interact.

Earth is a giant magnet. Magnets all over the planet line up with Earth's magnetic field. The north pole of every free-rotating magnet points to the magnetic north pole of Earth. That's why a compass is a good tool for keeping track of direction. A compass needle will point toward Earth's magnetic north pole.

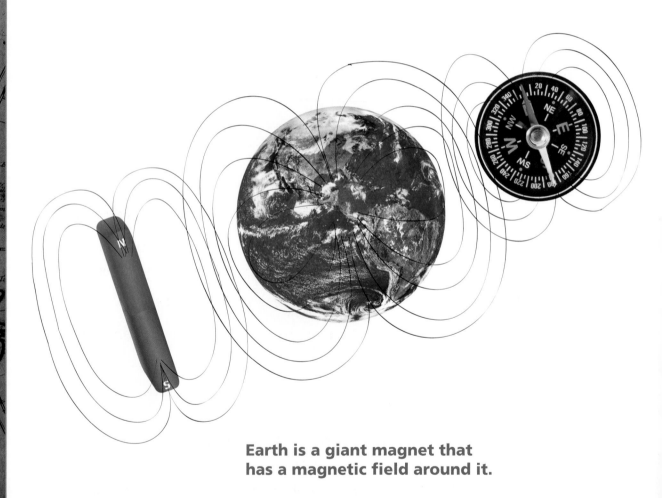

Earth is a giant magnet that has a magnetic field around it.

Make Another Homemade Compass

You can make another simple compass by turning a sewing needle into a permanent magnet. Here's how to do it.

What You Need

- 1 Steel sewing needle
- 1 Permanent magnet
- 1 Steel paper clip
- 1 Piece of plastic foam or cork
- 1 ½-liter container or cup
- Water
- Thread

What You Do

1 Tie one end of the thread to the paper clip.

2 Tie the other end of the thread around the piece of plastic foam or cork.

3 Using a permanent magnet, rub the sewing needle several times in one direction. Now the needle has two poles, just like every magnet.

4 Push the needle through the piece of plastic foam or cork.

5 Put the needle-and-paper clip system in the center of the container of water.

What Happens

The needle will float in the cup of water and rotate to line up with Earth's magnetic field. The needle is a compass!

The paper clip acts as an anchor so that the needle can freely rotate and won't get stuck on the side of the container.

Electricity Creates Magnetism

Oersted's Discovery

Hans Christian Oersted (1777–1851) was born in Denmark. As a young child, he lived with a family in Germany. He learned to read and became skilled in math, even though he never attended a school. By the time he was 11 years old, Oersted returned home and worked in his father's pharmacy. There he learned the basics of chemistry.

By the time he was 17 years old, Oersted passed the exam to enter the University of Copenhagen in Denmark. He studied chemistry, astronomy, physics, and math. After he graduated, he became a physics professor at the university.

Hans Christian Oersted

Like many scientists of his time, Oersted was fascinated with the discovery made by Alessandro Volta (1745–1827) in 1800. Volta invented the battery, which was the first source of electric current. The D-cell we use today is a direct result of Volta's discovery. Oersted conducted lots of experiments with electric current.

In 1820, Oersted was giving a lecture demonstrating that electric current makes wires hot. When he closed the electric circuit, a compass needle that happened to be on the lecture table rotated. Some people think he was planning to show the relationship between electric current and magnetism that day. Others think it was just a lucky accident. We will never know for sure.

Here's what might have happened. Oersted had a thin wire connected to a battery and a switch. A compass needle was right under one of the wires forming the circuit.

When the circuit was closed to deliver electric current to the thin wire, the compass needle rotated.

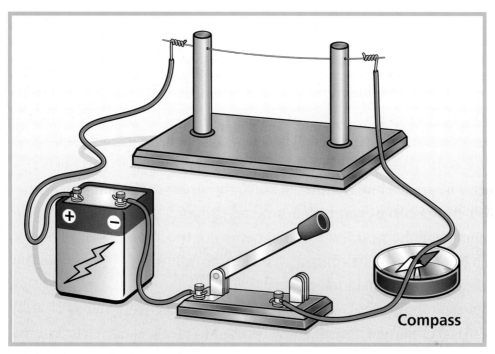

Oersted's demonstration with the switch open

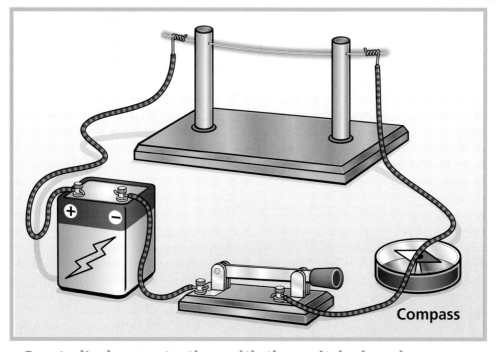

Oersted's demonstration with the switch closed

Oersted must have been excited, but he didn't announce his discovery at that time. He conducted more experiments. Four months later he wrote a report of his finding. His conclusion was that a flow of electric current produces a magnetic field.

This important discovery resulted in hundreds of inventions in the years that followed. One was the **electromagnet**, a magnet that can be turned on and off.

Magnetic Fields

The magnetic field around a wire that has electric current flowing through it is not very strong. If you can put two magnetic fields together, the magnetism will be stronger. That's what happens when a wire is wound into a **coil**. The magnetic field around each coil adds to the fields from other coils. The greater the number of coils, the stronger the total magnetic field is.

When the coil is wrapped around an iron or a steel **core**, like a rivet, the strong magnetic field induces magnetism in the steel. The steel becomes a temporary magnet as long as the current is flowing. And, with a flip of the switch, the magnetism turns off.

It wasn't long after Oersted's discovery that Michael Faraday (1791–1867) discovered that magnetism could be used to create electric current. From that time on, there was no doubting that one force was responsible for both magnetism and electricity. That force is the electromagnetic force.

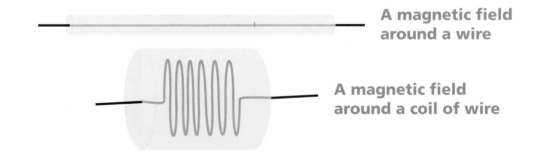

A magnetic field around a wire

A magnetic field around a coil of wire

Thinking about Electromagnetism

1. What was Oersted's historic discovery?

2. How does an electromagnet work?

3. Why did Oersted's compass needle rotate when he ran electric current through the thin wire?

Using Magnetic Fields

The wire you have been using to make circuits is made of copper. Copper wire is not magnetic. There is no magnetic field around a copper wire. You can confirm or prove this by bringing a compass close to a copper wire. The compass needle does not move.

Things change when you connect a copper wire to a source of electricity, such as a D-cell. While electric current is flowing through the wire, a magnetic field surrounds it. When you bring a compass close to the wire, the compass needle will rotate. When the circuit is broken, the magnetic field disappears, and the compass needle points north again.

The fact that a current produces a magnetic field can be used to make a very useful device. This device is an electromagnet. An electromagnet is a magnet that can be turned on and off with the flip of a switch.

The magnetic field surrounding a current-carrying wire is not very strong. The strength of the field can be increased by coiling the wire. When two coils are next to each other, the magnetic fields add together. This makes the magnetic field stronger.

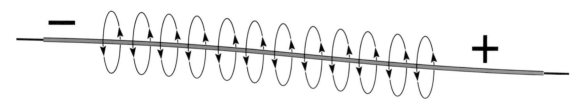

A magnetic field around a current-carrying wire

A magnetic field around a coil of current-carrying wire

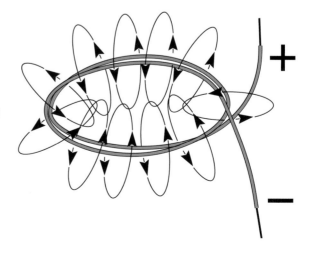

As you add more coils, the magnetic field gets stronger and stronger. If you put an iron core, like a rivet, in the center of the coils, the strong magnetic field will induce magnetism in the iron. That is how electromagnets are made.

There are several ways to make an electromagnet stronger. More coils of wire is one way that you have already learned about. The more wire you wrap around the core, the stronger the magnetism.

Another way to make an electromagnet stronger is to increase the amount of electric current flowing through the wire. You used one D-cell. Two D-cells increase the magnetism a lot. What if you had ten D-cells in series? Or 100 D-cells in series? Now we're talking about some strong magnetism.

A third way to make an electromagnet stronger is to wrap thicker wire around the core. Thicker wire can conduct more current. The thicker the wire is, the stronger the magnetic field, and the stronger the electromagnet.

There are a couple of other things you always need to think about. The wire must be insulated. (Do you know why?) And the coils of wire must all be wound around the core in the same direction. It doesn't matter which direction as long as they all go in the *same* direction.

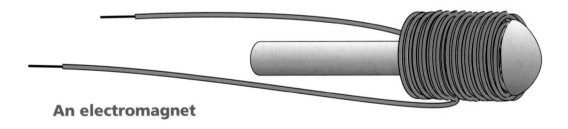

An electromagnet

Discuss Your Ideas

1. What is one way to tell if a wire has current flowing through it?

2. How do you make an electromagnet?

3. How can you make an electromagnet stronger?

Electromagnets Everywhere

Motors

A motor that runs on a D-cell is a direct-current motor. A direct-current motor has two main parts. They are permanent magnets and electromagnets.

A simple motor is like a tin can with two permanent magnets stuck inside. In the center of the can, there is a shaft that has two or more iron cores attached. A lot of wire is wrapped around each of the cores to make electromagnets.

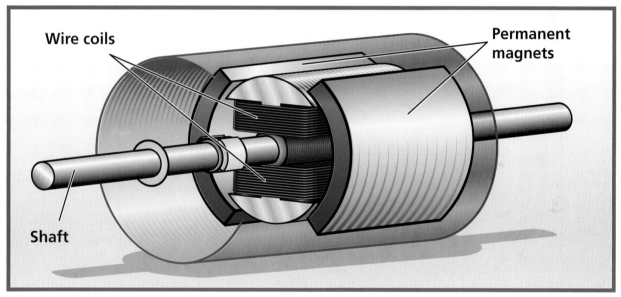

Wire coils

Permanent magnets

Shaft

The parts of a simple motor

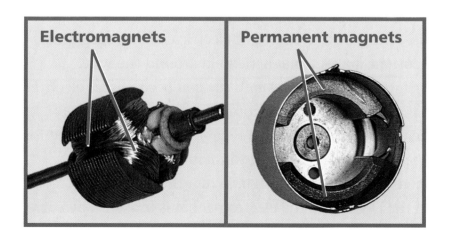

Electromagnets

Permanent magnets

Imagine taking the permanent magnets and shaft out of the can. Take off all but one of the wire coils. The simplified motor would look like the one below.

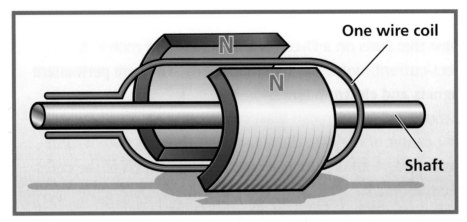

A simplified motor with two permanent magnets and only one wire coil

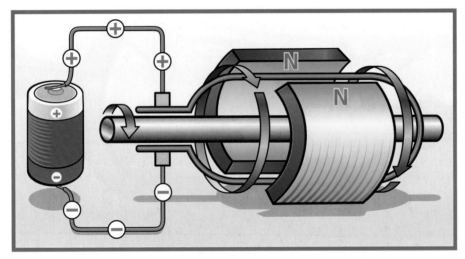

The wire coil becomes an electromagnet when it is connected to a D-cell.

Connect the wire to a source of electric current, such as a D-cell. The flow of current makes a magnetic field around the wire.

When current flows in the coil, the coil becomes an electromagnet. The magnetic field of the electromagnet is repelled by the fields of the permanent magnets. (Like poles repel.) This pushes the electromagnet away. The push causes the shaft to rotate.

But there is more to the design. When the shaft rotates, contact between the D-cell and the electromagnet is broken. The current stops flowing in the coil. The **electromagnetism** stops briefly.

As the shaft rotates a little farther, contact is made again. This creates the electromagnetic field in the coil again. The coil gets another magnetic push to keep the shaft turning.

The shaft gets hundreds of little magnetic pushes every second. The motor uses electric current and magnetism to produce motion.

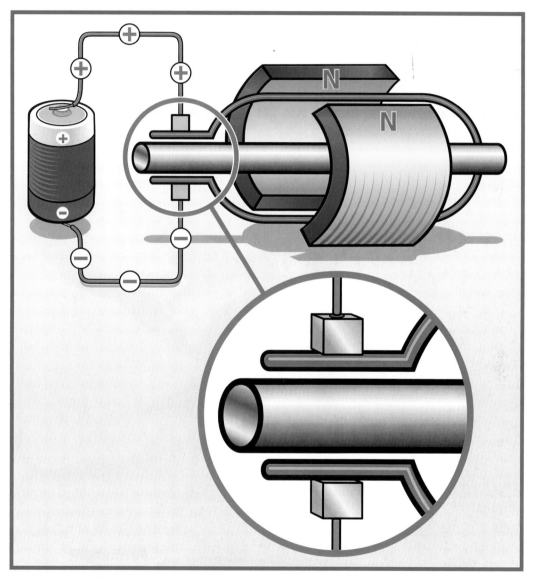

As the shaft rotates, contact between the D-cell and the coil is broken. As the shaft rotates a little farther, contact is made again.

Generators

You know how to use electricity to make a flashlight work. You put some D-cells in it. But where does the electricity in your home come from? It does not come from batteries. The electricity in your home comes from a generator. Generators use motion to produce electric current.

A generator has two main parts. They are permanent magnets and wire coils. They are the same parts found in a motor. A direct-current generator is a motor running in reverse.

When you put electricity into a motor, you get motion. When you rotate the shaft of the motor, you get electricity. Here's how it works.

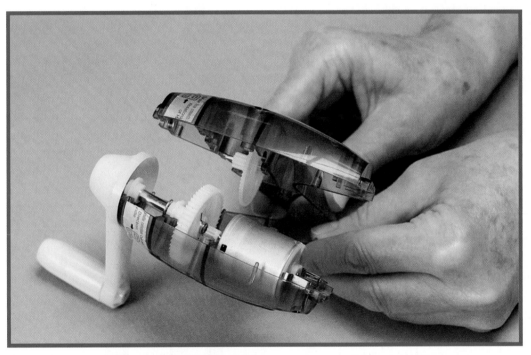

A simple hand-operated generator

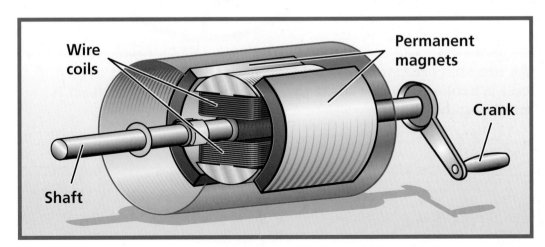

One wire coil

When a wire coil turns in a magnetic field, an electric current is created in the wire.

Contacts

Electricity flows from the generator when contact is made with the end of the wire coil.

There is a magnetic field between the poles of the permanent magnets. Set a shaft with wire coils between the magnets. Take off all but one of the wire coils. The simplified generator would look like the diagram above.

When a wire passes through a magnetic field, an electric current is created in the wire. If you rotate a wire coil in a magnetic field, it will pass through the field hundreds of times in a second. This makes a continuous flow of electric current.

What turns the wire coil in the magnetic field? It could be many things. A windmill can be attached to the shaft. Wind can then rotate the coil. Water flowing downhill, steam from a boiler, or a gas engine can also be used to rotate the coil.

As a last resort, you can put a crank on the end of a generator shaft. A crank lets you turn the wire coil by hand to generate a little electricity.

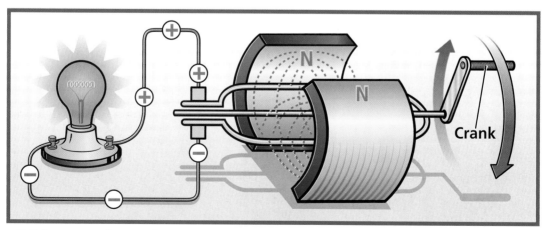

Crank

With a crank on the end of the shaft, you can generate electricity by hand.

Doorbells

When you press a doorbell button, it completes a circuit, and you hear the loud "r-r-r-r-r-ing." The button is a switch. The sound comes from a bell being hit by a little hammer. The hammer is called a striker. But the striker does not hit the bell just once. It hits the bell dozens of times a second. How does a doorbell work?

A doorbell has a number of components that make a circuit. It has an electricity source, a doorbell button (the switch), a box terminal, a striker terminal, a movable striker, an electromagnet, and a bell connected to the box. Find these parts in the drawing below.

The box terminal is attached to the box holding the bell. The striker terminal is attached to the striker. When no electric current is flowing through the circuit, the box terminal and striker terminal touch.

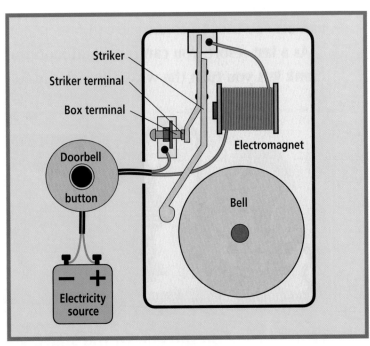

Components in a doorbell

When you press the doorbell button, current flows from the electricity source through the contact between the box terminal and the striker terminal. Then current flows through the striker, to the electromagnet, and back to the electricity source. The dotted lines show the circuit.

Current flows in the doorbell circuit and turns on the electromagnet.

When current flows, it activates the electromagnet. The electromagnet attracts the steel striker, and two things happen. The striker hits the bell. And the striker terminal pulls away from the box terminal. This breaks the circuit. When the circuit is broken, the magnetism goes away. The striker returns to its starting position. That brings the terminals back together so that the circuit is complete. The whole process starts over. The bell ding-ding-dings as long as you hold the button.

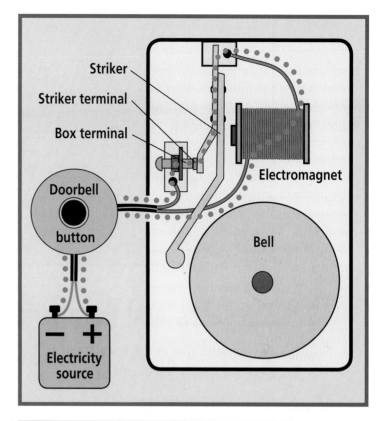

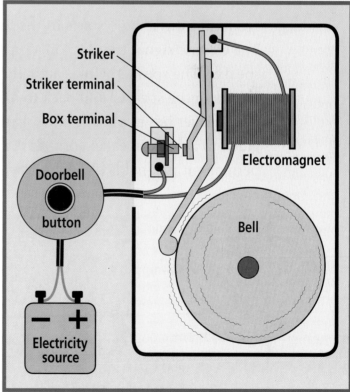

The electromagnet attracts the striker. The bell rings. The striker terminal gets pulled away from the box terminal. The circuit is broken. The magnetism goes away.

Speakers and Earphones

Speakers use electric current to produce motion. You might have felt the vibrations from speakers when music is playing. How does a speaker produce motion from an electric current?

Sound from a radio starts in the form of electricity. The electric signal travels along a speaker wire to the speaker. The speaker wire is actually two wires. The sound signal travels to the speaker and back to make a complete circuit.

A speaker has two main parts. Part 1 is a coil of insulated wire glued to a paper speaker cone. Part 2 is a permanent magnet. The permanent magnet is placed in the center of the cone.

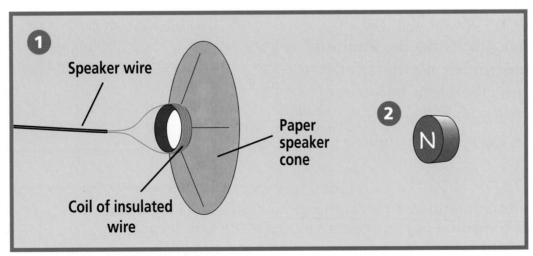

1 Speaker wire

Paper speaker cone

2 N

Coil of insulated wire

1. Speaker cone and coil   **2. Permanent magnet**

When a pulse of electricity flows through the coil, it creates a magnetic field. The magnetic field of the coil and the magnetic field of the permanent magnet repel. Because the coil is glued to the speaker cone, the cone moves. The movement of the cone pushes on air. This produces sound.

If you put two little speakers in the ear cups of a headset, you have a pair of earphones. If you put two tiny speakers into your ears, you have earbuds.

What if current flowing in a wire did not produce a magnetic field? How would your life be different without electromagnets?

Thinking about Electric Devices

1. How does a motor work?

2. How does a doorbell make a continuous ring?

3. How does a speaker work?

Morse Gets Clicking

Imagine what it would be like without cell phones and computers. How would you keep in touch with your friends? That's the way it was 170 years ago. There were no phones, radios, or televisions. There were no computers for e-mail. People wrote letters. It could take weeks or months to receive a letter and respond to it.

Samuel Finley Breese Morse

In 1820, Hans Christian Oersted discovered that a wire carrying an electric current produced a magnetic field. In 1825, the electromagnet was invented. People knew that electricity moved fast through wires. Could words be changed into electricity to speed up communication?

Samuel Finley Breese Morse (1791–1872) had an idea about how to make electricity "speak." In 1835, Morse used an electromagnet, a switch, and long wires to send a long-distance message. The switch (called a **key**) and a battery were in one location. Long wires ran to an electromagnet far away. When the key was pressed to complete the circuit, the electromagnet attracted a piece of steel with a loud click. Those first clicks announced the invention of the **telegraph**.

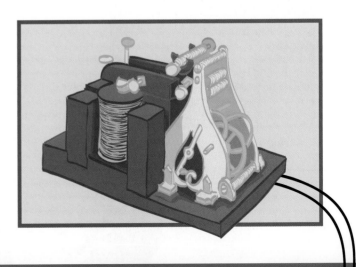

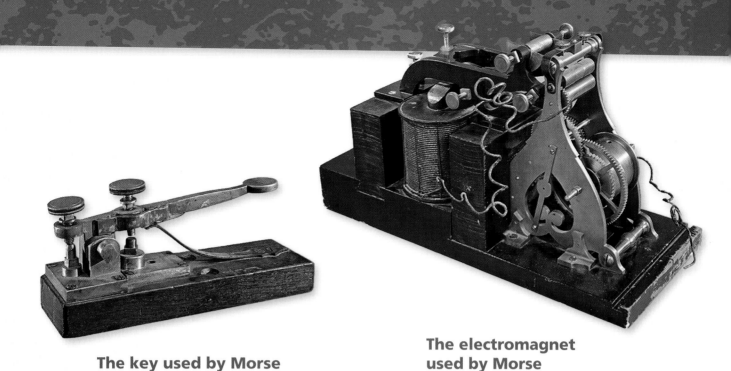

The key used by Morse

The electromagnet used by Morse

Morse knew his telegraph could change the way people communicated. But he had trouble finding others who agreed. Finally, 8 years later, he got a $30,000 grant to set up a telegraph line from the railroad station in Baltimore, Maryland, to the Supreme Court building in Washington, DC. The next year, in 1844, the first message traveled the 65 kilometers (km) between the two cities in a fraction of a second. In moments, the return message reached Morse. The telegraph was a success.

Morse's telegraph was not very different from the one you made in class. The key was similar to your switch. The batteries were stronger. The electromagnet had more winds of wire to make it stronger.

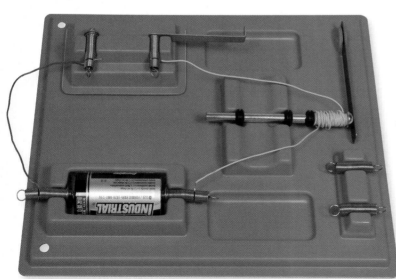

The telegraph you made in class

Putting the Message into Words

There's one more important part to this story. Remember, the telegraph didn't send words, it just sent clicks. How can you make words out of clicks? Morse needed a **code** to translate clicks into words.

The first code Morse tried didn't work very well. The receiver had a pen attached to the electromagnet. When the circuit was closed, the pen moved and made a mark on a roll of paper. The dots and squiggles were too hard to decode.

Next, Morse developed a code where clicks stood for words. For example, one click might be the word *you*. Two rapid clicks might be the word *today*. Two widely spaced clicks might be the word *now*, and so on. The code book was huge. It took a long time to look up the words after a set of clicks was received.

The code that worked was developed by Morse and his partner, Alfred Vail (1807–1859), in 1838. It used short and long sounds, called dots and dashes, to stand for letters of the alphabet. You might wonder how Morse got his telegraph to make short and long clicks. He didn't. The short and long "sounds" were actually the pauses between the clicks. The telegraph receiver heard the short and long pauses between clicks as the dots and dashes.

A modern high-speed key uses code to send messages in dots and dashes.

After the first successful message was sent in 1844 from Baltimore, Maryland to Washington, DC, the telegraph became popular. Every major city had a central telegraph office. Businesses, newspaper offices, and governments depended on the telegraph for the fast delivery of news. The telegraph changed how the world communicated.

The code was modernized in 1848 and became known as the International Morse Code. Even though advances in communication soon made Morse Code and the telegraph obsolete for everyday use, the code still has some important uses more than 160 years after its invention. For years, airplane pilots and ship captains have been required to learn the code so they could communicate and respond to coded messages sent by automatic location identifiers at sea and on land. Navy ships can also communicate with one another silently by sending Morse Code using flashing lights, if they need to avoid detection by enemy ships. The international distress signal SOS is communicated even today by Morse Code and is understood all over the world. Dit–dit–dit; dah–dah–dah; dit–dit–dit is the sound that declares "I need assistance immediately." The SOS message can be sent in a number of ways, such as by flashing a mirror, by turning a flashlight on and off, or by keying a radio on and off.

A .–	J .– – –	S ...	1 .– – – –
B –...	K –.–	T –	2 ..– – –
C –.–.	L .–..	U ..–	3 ...– –
D –..	M – –	V ...–	4 ....–
E .	N –.	W .– –	5 .....
F ..–.	O – – –	X –..–	6 –....
G – –.	P .– –.	Y –.– –	7 – –...
H	Q – –.–	Z – –..	8 – – –..
I ..	R .–.		9 – – – –.
			0 – – – – –

Period .–.–.–   **Apostrophe** .– – – –.   **Right parenthesis** –.– –.–

Comma – –..– –   **Hyphen** –....–   **Left parenthesis** –.– –.

Colon – – –...   **Slash** –..–.   **Quotation marks** .–..–.

Question mark ..– –..   **At sign (@)** .– –.–.

Here is the present-day International Morse Code.

Beyond the Telegraph

The telegraph system developed by Morse was very successful. It did have one limitation. Only one message at a time could be sent over a telegraph line. With very few lines, only a few messages could be sent in a day. Several people tried to develop a way to send multiple messages along the line at one time. One of the people working on this problem was Alexander Graham Bell (1847–1922).

Alexander Graham Bell

Bell's idea was that pulses of electricity of different frequencies could share the line at the same time. In his efforts to create electrical pulses of different frequency, Bell applied his knowledge of music to the problem. He was experimenting with methods to send pulses of electricity at high pitches and low pitches at the same time. What he needed was a device to produce electrical pulses in response to sounds of different frequencies.

In his efforts to develop such a device, he realized that the human voice could produce the range of frequencies he needed. The device that he developed produced electric currents of different intensities depending on the vibrations of the pitch of the note directed into the device. It occurred to Bell that if he could develop a second device to receive the pulses of electricity and convert them back into vibrations, he could transmit the human voice over a wire.

In 1876, Bell and his assistant Thomas Watson (1854–1934) successfully transmitted the human voice over a wire. The telephone was invented! The invention of the telephone also marked the beginning of the end for the telegraph. Speaking words was much more desirable than sending dots and dashes.

This model of Bell's first telephone is a duplicate of the one through which speech sounds were first transmitted electrically.

Guglielmo Marconi and his invention

Communication Leaves the Wires

In 1888, Heinrich Hertz (1857–1894) was demonstrating that electric impulses could be transmitted through the air. He developed a device that generated electric sparks. This device created electromagnetic radiation that traveled out in all directions. This radiation is now called radio waves.

A young Italian inventor, Guglielmo Marconi (1874–1937), took an interest in Hertz's discovery. Marconi began developing an apparatus to send messages over long distances without wire. He assembled a transmitter, which was little more than a spark generator. He added an antenna to capture the electromagnetic pulses from the air. And he added a receiver to turn the electromagnetic pulses into sounds that could be interpreted. Before long, Marconi's system was able to send and receive messages over a distance of more than 1 km. By 1897, Marconi was able to send messages more than 6 km. In 1902, he successfully sent a wireless message across the Atlantic Ocean. By 1915, a new technology allowed audio transmissions. Now the airwaves could be used for radio transmissions of human voice and entertainment programming.

Radio Meets the Telephone

The familiar cellular telephone carried by just about everyone these days was first conceived in 1947. At that time, mobile telephones were installed in cars because the phones needed a large, reliable supply of electricity. The phones were limited because they only worked near a transmission tower. The tower would receive the signal from the phone and send it to another phone, but both phones had to be near the same transmission tower.

The cellular idea was to develop a network of small service units, called cells, all over the country. A cellular phone user would always be in one cell or another. The phone user has only to communicate with the transmission tower in the small cell. The cell transmitter then communicates with the cell tower in the next cell. The signal gets relayed from cell to cell until it arrives at the receiver phone. It might be a block away or across the country.

Cell phones use radio waves to transmit signals. So your phone is a radio transmitter and radio receiver. Instead of music or sports, your radio receiver sends and receives only the conversation you want to have with your friend. It's as if you have your own personal radio station. Text messages travel the same way, but letters of the alphabet are sent instead of spoken words.

Static Electricity

Did you ever reach for a doorknob and get a shock? Zap! Why does that happen? It's because of **static electricity**.

Static electricity starts with atoms. Atoms are the small particles that everything is made of. Atoms have a nucleus in the center. The nucleus has a **positive charge**. Atoms also have tiny **electrons** moving around the nucleus. Electrons have a **negative charge**.

Most of the time, the number of positive charges in the nucleus is the same as the number of negative charges moving around the nucleus. When the positive and negative charges are equal, the atom is **electrically neutral**.

Electron Transfer

Sometimes an electron can leave one atom and move to another atom. When this happens, one atom has gained an electron, and one has lost an electron. The atom that has gained an electron has a negative charge. The atom that lost an electron has a positive charge.

Rubbing two objects together can cause electrons to transfer from one object to the other. If you rub a wool sweater on a balloon, electrons transfer from the wool to the rubber. This transfer is how objects can get a static electric charge.

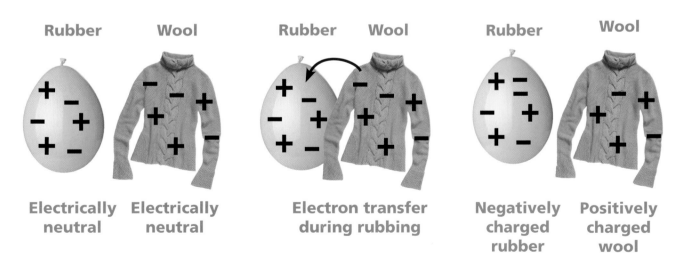

Rubber	Wool	Rubber	Wool	Rubber	Wool
Electrically neutral	Electrically neutral	Electron transfer during rubbing		Negatively charged rubber	Positively charged wool

The Balloon Experiment

When two balloons hang by threads from a single point, they come to rest touching each other. The balloons are electrically neutral.

When both balloons are rubbed on a wool sweater or someone's hair, the balloons push each other apart. Why?

During rubbing, electrons transfer from the sweater to the balloons. Both balloons get a negative charge. Like charges repel each other. That's why the balloons don't touch.

Do you think there is a charge on the wool sweater after rubbing the balloons? If so, is it a negative or positive charge? Remember, electrons transferred from the sweater to the balloons give the balloons a negative charge. That means the sweater lost electrons. The sweater has a positive charge.

What will happen when you bring the positively charged sweater close to the negatively charged balloons? The sweater will attract the balloons. The charge on the wool sweater is opposite to the charge on the balloons. Opposite charges attract.

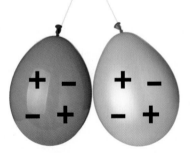

Two balloons hang side by side when they are electrically neutral.

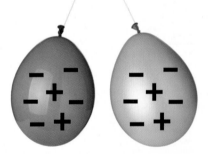

Two balloons repel each other when they have like charges.

Thinking about Static Electricity

1. Suppose you wear rubber-soled shoes and walk across a wool rug. What will happen when you reach toward a charged balloon hanging on a thread? Why do you think so?

2. Suppose you wear wool slippers and walk across a rubber floor. What will happen when you reach toward a charged balloon hanging on a thread? Why do you think so?

Light Interactions

Light is evidence of energy. Light comes from a **light source**. The Sun is a light source. A lightbulb is a light source. A flame is a light source. Anything that makes light is a light source. Can you think of any other light sources?

Light travels in rays. Light rays travel from a light source in straight lines in all directions. Light rays don't curve around things. They just travel straight. And they will travel forever if they don't run into anything.

A candle is a small light source. It is safe to look at a candle. When light rays from a candle flame enter your eyes, you can see the flame. If light rays from the flame don't enter your eyes, you can't see the flame. You can only see something if light travels from it into your eyes.

A candle flame is a light source.

Reflected Light

Can you see the picture of a candle on this page? If you can, light must be traveling from the picture into your eyes. But the picture of the candle is not making light. Where is the light coming from?

Look around. Are the lights on in the room? Is there a window where light can come in? That's where the light is coming from. Light from lightbulbs and the Sun is traveling to the candle picture. Then the light bounces off the picture into your eyes. Light bouncing off a surface is called **reflection**.

A lightbulb is a light source. Light rays travel from the source in straight lines. Some of the light rays strike the candle picture. The light rays reflect off the picture. When the light reflects, it changes direction. But it still travels in a straight line. When light from the candle picture reflects into your eyes, you see the picture.

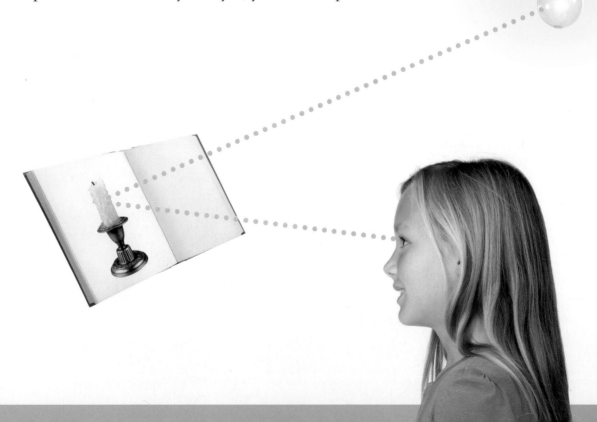

A duck and its reflection in a mirror   A mirror can show what is behind you.

Mirrors

What do you see when you look in a **mirror**? Often you see yourself, but not always. You can hold a mirror to see things in other directions. In fact, if you hold a mirror just right, you can see objects behind you. It's like having eyes in the back of your head.

Mirrors are shiny surfaces that reflect light. You can use a mirror to reflect light into your eyes. That's how you are able to see yourself in a mirror. That's how drivers can see what's going on behind them. And that's how sailors in submarines look around the ocean's surface. They use a device with two mirrors called a periscope.

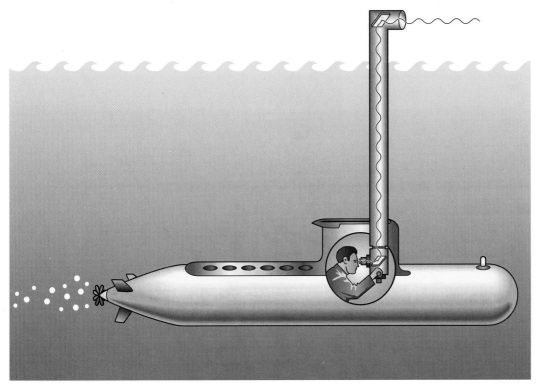

A submarine periscope can show what is above you.

Mirrors can also be used to change the direction of a beam of light. Mirrors can direct light around an object.

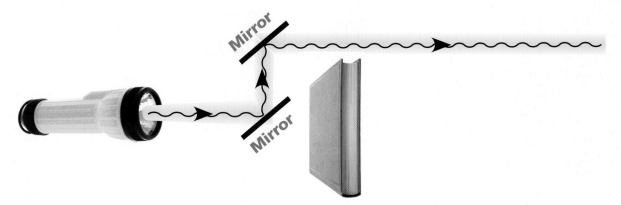

Two mirrors can direct light back to the source.

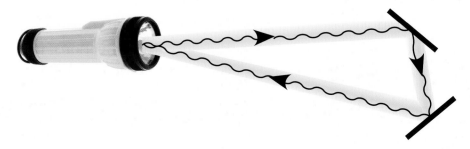

With four mirrors, you can make it look like light shines through a solid object.

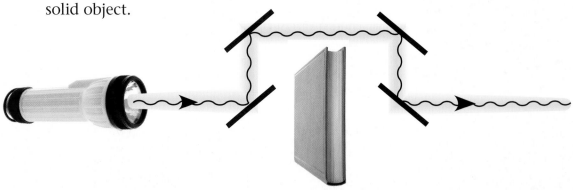

Two mirrors can be used to reflect light in two directions at the same time.

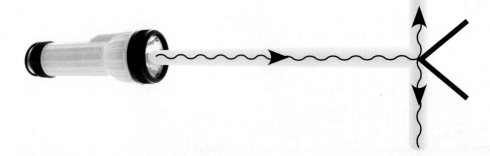

Other smooth, flat surfaces can act like mirrors. Sometimes you can see your reflection in a glass window. The surface of a calm lake can reflect light, too.

Did you ever look at your reflection in a shiny spoon? Something funny happens. On the back of the spoon, you look tall and skinny. In the bowl of the spoon, you look small and upside down. Curved surfaces reflect light in interesting ways.

Spoon reflections are fun.   **Glass reflects light to show this cat's reflection.**

77

Refraction

Light travels at different speeds. It moves very fast through air, but it moves slowly through things that are more dense than air. The more dense the substance, the more slowly light travels through it. That's why a light ray moving through water, plastic, or glass seems to bend. These materials are more dense than air. We call this bending of light rays **refraction**.

A hot surface can change the density of air just above it. When that happens, light is refracted where the hot air meets a layer of cooler air. The refraction makes you think you see something that is not there. This illusion is called a mirage. On some days, you might see a mirage that looks like a pool of water above a hot, paved road that is completely dry.

Straws in a glass of water look broken because of refraction.

Reflecting on Light

1. What must happen for you to see an object?

2. What happens when light reflects?

3. What kinds of surfaces reflect light?

4. What can you use a mirror for?

5. What happens when light refracts?

Throw a Little Light on Sight!

Sara's class was on a field trip at the Lawrence Hall of Science.

They were studying light. Sara was excited when she saw an exhibit called "Throw a Little Light on Sight!"

A helper was standing by the open door.

Would you like to come in and learn about light and vision?

Yes, we are studying light in class.

Good! Let's go into this dark room. It has no windows and no lights. I am closing the door.

What do you see?

I can't see anything.

I am putting two objects in front of you on the table. We will wait 5 minutes.

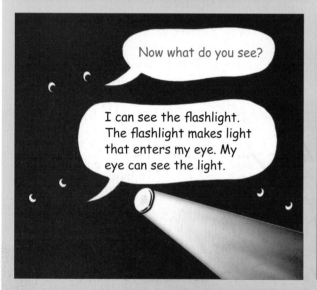

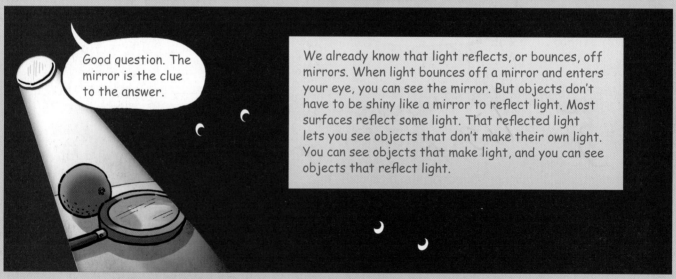

Different surfaces reflect different colors of light. The color of light shining on a surface affects how it looks. It is a fascinating part of the study of light.

Later That Week. . .

Sara tried some experiments at home. She got some clear blue plastic, some green plastic, and some yellow plastic. With these she could shine blue, green, and yellow light with her flashlight. With an orange and a lime from the kitchen, she was ready.

When she shined blue light on the orange and lime, they both looked black.

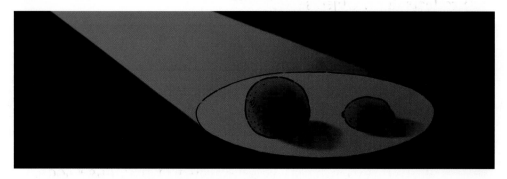

When she shined green light, the orange looked black and the lime looked green.

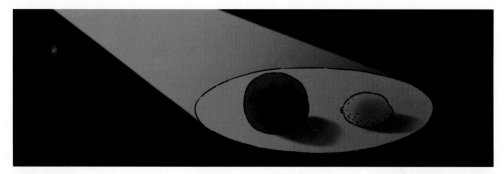

And when she shined yellow light, the orange looked yellow and the lime looked black.

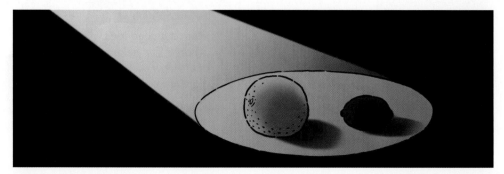

Sara remembered something she learned in class. White light is all colors of light mixed together. White light has yellow light, red light, green light, and all the other colors. A triangle of glass called a prism can separate the colors.

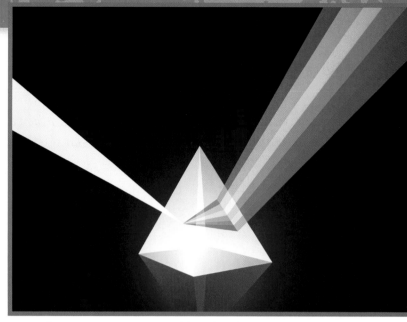

A prism separates white light into the colors of the rainbow.

When white light strikes an object, the object **absorbs** light or it reflects light. Objects like paper, snow, and cotton reflect almost all the light that hits them. That's why they look white. Objects like charcoal, pencil lead, and truck tires absorb almost all the light that hits them. That's why they look black.

When white light hits an orange, almost all the colors of light are absorbed except orange. Orange is reflected. That's why oranges look orange. All of a sudden Sara knew why the orange looked black when blue light shined on it. There was no orange light for the orange to reflect.

Sara thought, "I bet I can predict the appearance of a red apple in green light. And what if I had some red plastic for my flashlight? I could predict the appearance of the lime in red light."

Thinking about Light

1. Why couldn't Sara see anything when she first went into the exhibit at the Lawrence Hall of Science?

2. Why did Sara's orange appear black in blue light?

3. Why did Sara's lime appear green in white light?

4. How will Sara's lime look in red light? Explain why.

More Light on the Subject

Energy from the Sun comes to Earth as light. Energy of batteries can produce light. Energy of fuel can produce light. Anything that produces light is a light source.

Light travels from a light source in rays. The rays travel in straight lines. A light ray will travel forever in a line unless it hits an object. When light hits an object, two things can happen. The object might absorb the light. Or the object might reflect the light.

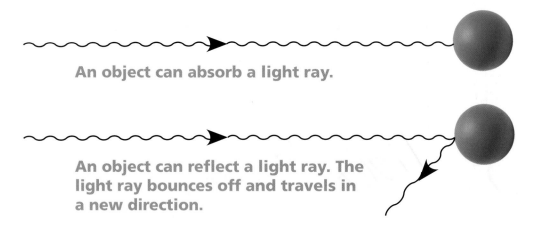

An object can absorb a light ray.

An object can reflect a light ray. The light ray bounces off and travels in a new direction.

Light that is absorbed is no longer light. This absorbed light can be changed into heat. Light that is reflected is still light. Reflected light bounces off an object and continues on its way. Reflected light travels in a new direction.

Mirrors

A mirror is a shiny surface. Light reflects from a mirror. A mirror can change the direction of light coming from an object. This **property** is useful when the light is coming from behind you. A mirror can change the direction of the light so that you can see what is going on behind you.

A flashlight makes a beam of light. A beam is millions of light rays. A mirror can be used to change the direction of a beam of light. Two mirrors can reflect light into a dark room down the hall and around the corner.

Most objects reflect light. That's how we are able to see them. Rays of light from a light source or rays of reflected light enter our eyes. When we see an object, we are actually seeing the light that travels from that object into our eyes. If no light enters our eyes, like when we are in a dark room, we see nothing.

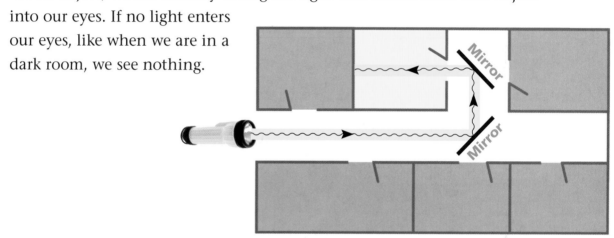

Drivers use a rear-view mirror to see behind them.

Seeing Color

Light from the Sun and from lightbulbs is called white light. But white light is really a mixture of all the colors of the rainbow. In fact, when you see a rainbow, you are seeing all the colors in white light. When conditions are right, tiny drops of water separate the colors.

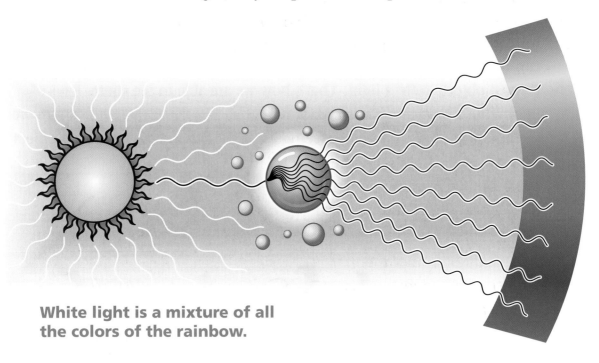

White light is a mixture of all the colors of the rainbow.

When white light strikes an object, some colors are absorbed and some are reflected. When white light shines on a red apple, all the colors of light except red are absorbed. Only red light is reflected. When the red light goes into your eyes, you see that the apple is red.

What will you see if you shine blue light on the same red apple? The apple will appear black. The blue light is absorbed by the apple. No light is reflected.

The color of light striking an object affects the way you see the object.

The apple appears red because it reflects only red light. Other colors of light are absorbed by the apple.

Into the Shadows

The day after the field trip to the Lawrence Hall of Science, Sara's teacher had more things for the class to think about. "When you leave school today, look for your **shadow**. When do you see your shadow? Is your shadow in front of you or behind you?"

After school, Sara went onto the playground with three friends. She saw her shadow right away. She looked closely as she lifted her arms up to her waist and then over her head. Her shadow did everything she did. And her shadow was right in front of her.

Sara then turned slowly in a circle. She thought that her shadow would stay in front of her as she turned. But it didn't. Sometimes her shadow was behind her. Sometimes it was on her right, and at other times it was on her left.

Sara realized that her shadow always pointed the same way. It didn't matter which way she faced. All the other shadows she saw pointed the same way as her shadow. "Why do shadows always point in the same direction?" she wondered.

Sara's friends can see their shadows.

When light from the Sun is blocked, you see a shadow.

It's because of the Sun! The Sun is in one place in the sky. The Sun shines on everything. Shadows always point away from the Sun. Sara observed that she was always between the Sun and her shadow.

Sara went down the street. Her shadow stayed with her. She crossed the street, and her shadow crossed with her. Then she moved under a big tree. Her shadow was gone!

"That's the answer to my teacher's first question," Sara thought. "I have a shadow when I am in the sunshine. Anything that sunlight can't shine through, like me, blocks the light from hitting the ground. A shadow is the dark place where no sunlight reaches the ground."

The next day Sara reported her discoveries.

"I saw lots of shadows yesterday. I had a shadow, my friends had shadows, and the swings and trees had shadows. I figured out that shadows are the dark places where light doesn't shine. When light is blocked by an object, the object has a shadow. Things that light can't shine through have shadows."

"Was your shadow in front of you?" asked Sara's teacher.

"My shadow was in front of me and in back of me. It always pointed the same way, so I could decide if I wanted it in front or in back. All I had to do was turn."

"Were you able to hide from your shadow?"

"The only way I could hide from my shadow was to get into the shadow of something else, like a tree. But as soon as I moved back into the sunlight, my shadow was right there again."

"I have one more question," said Sara's teacher with a twinkle in her eye. "Where does your shadow go at night?"

"Is that a trick question?" Sara smiled.

Science Safety Rules

1. Listen carefully to your teacher's instructions. Follow all directions. Ask questions if you don't know what to do.

2. Tell your teacher if you have any allergies.

3. Never put any materials in your mouth. Do not taste anything unless your teacher tells you to do so.

4. Never smell any unknown material. If your teacher tells you to smell something, wave your hand over the material to bring the smell toward your nose.

5. Do not touch your face, mouth, ears, eyes, or nose while working with chemicals, plants, or animals.

6. Always protect your eyes. Wear safety goggles when necessary. Tell your teacher if you wear contact lenses.

7. Always wash your hands with soap and warm water after handling chemicals, plants, or animals.

8. Never mix any chemicals unless your teacher tells you to do so.

9. Report all spills, accidents, and injuries to your teacher.

10. Treat animals with respect, caution, and consideration.

11. Clean up your work space after each investigation.

12. Act responsibly during all science activities.

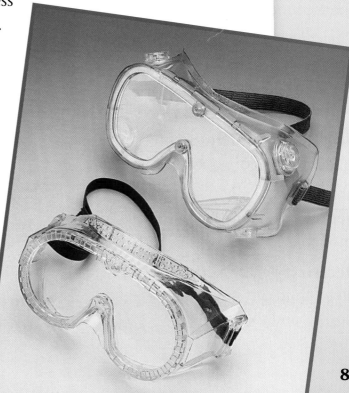

Glossary

absorb to take in or soak up

attract to pull toward each other

battery a source of stored chemical energy

circuit a pathway for the flow of electricity

closed circuit a complete circuit through which electricity flows

code a set of signals that represents letters or words for sending messages

coil a series of loops

compass an instrument that uses a free-rotating magnetic needle to show direction

complete circuit a circuit with all the necessary connections

component one item in a circuit

contact point the place in a circuit where connections are made to allow electricity to flow

core in an electromagnet, the material around which a coil of insulated wire is wound

electrically neutral an object that has equal numbers of positive and negative charges

electric current the flow of electricity through a conductor

electricity energy that flows through circuits and can produce heat, light, motion, and sound

electromagnet a piece of iron that becomes a temporary magnet when electricity flows through an insulated wire wrapped around it

electromagnetism a property of electric and magnetic fields that causes interactions with electric charges and currents

electron a tiny particle that has a negative charge and goes around the nucleus of an atom

energy the ability to do work

energy source a place where energy comes from, such as batteries, food, fuels, and the Sun

filament the material in a lightbulb (usually a thin wire) that makes light when heated by an electric current

force a push or a pull

fossil fuel the preserved remains of organisms that lived long ago and changed into oil, coal, and natural gas

generator a device that produces electricity from motion

heat observable evidence of energy

incomplete circuit a circuit that has a break in it

induced magnetism the influence of a magnetic field on a piece of iron, which makes the iron a temporary magnet

interact to act on and be acted upon by one or more objects

iron a metal that sticks to a magnet

key a switch that completes the circuit in a telegraph system

light observable evidence of energy

lightbulb a filament held by two stiff wires and surrounded by a clear glass globe

light source anything that makes light, such as the Sun, a lightbulb, or a flame

magnet an object that sticks to iron or steel

magnetic field an invisible field around a magnet

magnetism a property of certain kinds of materials that causes them to attract iron or steel

mirror a shiny surface that reflects light

motion observable evidence of energy

motor a device that produces motion from electricity

negative charge (–) the charge of an electron

north pole the end of a magnet that orients toward Earth's magnetic north pole

open circuit an incomplete circuit through which electricity will not flow

orient to position an object in a certain way

parallel circuit a circuit that has two or more pathways for current to flow

permanent magnet an object that sticks to iron

pole the end of a magnet

positive charge (+) the charge of an atom's nucleus

property something you can observe about an object or a material

reflection the bouncing of light rays off an object

refraction the bending of light rays

repel to push away from each other

series circuit a circuit that has only one pathway for current to flow

shadow the dark area behind an object that blocks light

sound observable evidence of energy

south pole the end of a magnet that orients toward Earth's magnetic south pole

static electricity positive and negative electric charges that don't move and are separated from each other

stored energy energy available for use

telegraph a device that uses an electromagnet to send coded messages by closing and opening an electric circuit

temporary magnet a piece of iron that behaves like a magnet only when it is surrounded by a magnetic field

vibration a quick back-and-forth movement

wire a metal or other solid substance through which electric current moves

Index